Mars Direct

Other Books by Robert Zubrin

Islands in the Sky
The Case for Mars: The Plan to Settle the Red Planet and Why We Must
Entering Space: Creating a Spacefaring Civilization
First Landing
Mars on Earth: Adventures of Space Pioneers in the High Arctic
The Holy Land
Benedict Arnold: A Drama of the American Revolution in Five Acts
Energy Victory: Winning the Mars on Terror by Breaking Free of Oil
How to Live on Mars: A Trusty Guidebook to Surviving and Thriving on the Red Planet
Merchants of Despair: Radical Environmentalists, Criminal Pseudo-Scientists, and the Fatal Cult of Antihumanism

Mars Direct

Space Exploration, the Red Planet, and the Human Future

BY ROBERT ZUBRIN

POLARIS BOOKS

Lakewood, Colorado

MARS DIRECT

POLARIS BOOKS
11111 W. 8TH AVE UNIT A
LAKEWOOD, CO 80215

FIRST PUBLISHED AS AN ELECTRONIC ORIGINAL BY
JEREMY P. TARCHER/PENGUIN
Penguin Group (USA) Inc., 375 Hudson Street, New York, New York 10014, USA

Manufactured in the United States of America

Layout and Cover by Carie Fay

Cover Picture Falcon Heavy, Courtesy of SpaceX

CONTENTS

There are no ends, limits, or walls that can bar us or ban us from the infinite multitude of things.

—Giordano Bruno, *On the Infinite Universe and Worlds*, 1584

When a population of organisms grows in a finite environment, sooner or later it will encounter a resource limit. This phenomenon, described by ecologists as reaching the "carrying capacity" of the environment, applies to bacteria on a culture dish, to fruit flies in a jar of agar, and to buffalo on a prairie. It must also apply to man on this finite planet.

—John P. Holdren and Paul R. Ehrlich, *Global Ecology*, 1971.

Part 1:

The Challenge of Mars

The Earth is not the only world. There are billions of other potential homes for life. And the first of these is now within reach.

The planet Mars is a world of towering mountains, vast deserts, polar ice fields, dry river channels, and spectacular deep canyons. Possessing a surface area equal to all the continents of the Earth put together, it orbits our sun at a distance about 50 percent greater than that of the Earth. This makes Mars a cold world, but not impossibly so. The average sunlight received at the Martian equator is about equal to that which shines upon Norway or Alaska. During the day at low Martian latitudes, the temperature frequently exceeds 50° F (10° C). At night, however, the thin Martian atmosphere does a poor job of retaining heat, and temperatures drop to –130° F (–90° C).

There is no liquid water on the surface of Mars today, but there was once, and our satellite probes show us its handiwork in the form of large networks of dried up riverbeds, dry lakes, and even the basin of a now-vacant northern Martian ocean. The water, however, is there—its surface reserves frozen as ice and permafrost and covered with dust, its deeper reservoirs still liquid, warmed by the planet's remaining sources of geothermal heat. There is as much water per square mile on Mars as there is on the continents of our home world.

Water is the staff of life, and the presence of large quantities of water on Mars marks it out as a potential home for a biosphere. On Earth, wherever we find liquid water, we find life. The evidence from our orbital images shows that there was liquid water on the surface of Mars for about a billion years of the planet's early history, a span roughly ten times as long as it took for life to appear in the Earth's fossil record after there was liquid water here. Thus if the conjecture is correct that life is a natural development from chemistry wherever one has liquid water and a sufficient period of time, then life should have appeared on Mars. Fossils recording its history may be there for us to find.

Life may have lost its foothold on the planet's surface, with the loss of the juvenile Mars's early thick carbon dioxide atmosphere and its associated greenhouse warming capability. But our space probes show that liquid water has gushed out from the Red Planet's subsurface within the past few million years, and probably within the past decade. In either case, effectively, the geologic present. This means that refuges for retreating Martian life may still exist. If we go there and drill, we could find them, and in finding them determine whether life as we know it on Earth is the pattern for all life everywhere or whether we are just one example of a much vaster and more varied tapestry. Mars is thus the Rosetta stone that will reveal to us the nature of life and its place within the cosmic order.

The New World

But Mars is more than just an object of scientific inquiry. It is a world capable of sustaining not only an ancient native microbial ecology, but a new immigrant branch of human civilization. For the Red Planet's resources go well beyond its possession of water. It has carbon in abundance as well, present both in the carbon dioxide that composes the majority of its atmosphere and in carbonates in its surface material. It has nitrogen, too; nitrogen is the leading minority gas in Mars's air and almost certainly exists as nitrates in the soil as well. Thus between the water, carbon dioxide, and nitrogen, we have all four of the primary elements of life (carbon, nitrogen, oxygen, and hydrogen). Calcium, phosphorus, and sulfur—the key secondary elements of life—are present in abundance as well. (In contrast, with the exception of oxides bound in rock, or ultra-cold condensations found in permanently shadowed polar craters, all of these are either rare or virtually absent on the Earth's Moon.)

In addition, all the elements of industry, such as iron, titanium, nickel, zinc, silicon, aluminum, and copper, are available on Mars, and the planet has had a complex geological history involving volcanism and hydrological action that has allowed for the concentration of geochemical rare elements into usable concentrated mineral ore. Mars's day-night cycle is 24.6 hours long, nearly the same as the Earth, which is not only pleasant for humans, but more importantly, makes it fully suitable for growing plants in outdoor greenhouses using natural sunlight. The planet's geothermal heat, which currently may sustain the habitats for scientifically fascinating native microbes, can also be used to provide both plentiful liquid water and power for human Mars settlements.

In a way that simply is not true of the Earth's Moon, the asteroids, or any other extraterrestrial destination in our solar system, Mars is the New World. If we can go there and develop the craft that allows us to transform its native resources into usable materials—transforming its carbon dioxide and water into fuel and oxygen, using its water and soil and sunlight to grow plants, extracting geothermal power from its subsurface, using its collection of solid resources to produce bricks, ceramics, glasses, plastics, and metals, making our way up the ladder of craftsmanship to make wires, tubes, clothes, tankage, and habitats—then we can create the technological underpinnings for not only a new branch, but a new *type* of human society.

Because it is the closest world that can support settlement, Mars poses a critical test for the human race. How well we handle it will determine whether we remain a single planet–constrained species or become spacefarers with the whole universe open before us.

Part 2:

How It Can Be Done

Mars is the New World. Someday millions of people will live there. What language will they speak? What values and traditions will they cherish, to spread from there as humanity continues to move out into the solar system and beyond? When they look back on our time, will any of our other actions compare in value to what we do today to bring their society into being? Today, we have the opportunity to be the founders, the parents and shapers of a new and dynamic branch of the human family, and by so doing, put our stamp upon the future. It is a privilege not to be disdained lightly.

Many people believe that a human mission to Mars is a venture for the far future, a task for "the next generation." Such a point of view has no basis in fact.

On the contrary, the United States has in hand, today, all the technologies required for undertaking an aggressive, continuing program of human Mars exploration, with the first piloted mission reaching the Red Planet within a decade. We do not need to build giant spaceships embodying futuristic technologies in order to go to Mars. We do not need to build a lunar base, a grander space station, or seek any other way to mark time for further decades. We can reach the Red Planet with relatively small spacecraft launched directly to Mars by boosters embodying comparable technology as that which carried astronauts to the Moon almost a half century ago. The key to success comes from following a travel-light-and-live-off-the-land strategy that has served explorers well over the previous centuries that humanity has wandered and searched the globe. A plan that approaches human missions to the Red

Planet in this way is known as the Mars Direct approach. Here's how it would work.

The Mission

At an early launch opportunity (for example, 2022), a single heavy lift booster with a capability equal to that of the Saturn V used during the Apollo program is launched off Cape Canaveral and uses its upper stage to throw a forty-metric ton (or "tonne")unmanned payload onto a trajectory to Mars. Arriving at Mars eight months later, it uses friction between its aeroshield and Mars's atmosphere to brake itself into orbit around Mars, and then lands with the help of a parachute. This payload is the Earth Return Vehicle (ERV), and it flies out to Mars with its two methane/oxygen-driven rocket propulsion stages unfueled. It also has with it six tonnes of liquid hydrogen cargo, a hundred-kilowatt nuclear reactor mounted in the back of a methane/oxygen-driven light truck, a small set of compressors and automated chemical processing unit, and a few small scientific rovers.

As soon as landing is accomplished, the truck is telerobotically driven a few hundred meters away from the site, and the reactor is deployed to provide power to the compressors and chemical processing unit. The hydrogen brought from Earth can be quickly reacted with the Martian atmosphere, which is 95 percent carbon dioxide gas (CO_2), to produce methane and water, and this eliminates the need for long-term storage of cryogenic hydrogen on the planet's surface.

The methane so produced is liquefied and stored, while the water is electrolyzed to produce oxygen, which is stored, and hydrogen, which is recycled through the methanator. Ultimately these two reactions (methanation and water electrolysis) produce twenty-four tonnes of methane and forty-eight tonnes of oxygen. Since this is not enough oxygen to burn the methane at its optimal mixture ratio, an additional thirty-six tonnes of oxygen is produced via direct dissociation of Martian CO_2. The entire process takes ten months, at the conclusion of which a total of 108 tonnes of methane/oxygen bipropellant will have been generated. This represents a leverage of eighteen to one of Martian propellant produced compared to the hydrogen brought from Earth needed to create it. Ninety-six tonnes of the bipropellant will be used to fuel the ERV, while twelve tonnes are available to support the use of high-powered, chemically fueled, long-range ground vehicles. Large additional stockpiles of oxygen can also be produced, both for breathing and for turning into water by combination with hydrogen brought from Earth. Since water is 89 percent oxygen (by weight), and since the larger part of most foodstuffs is water, this greatly reduces the amount of life-support consumables that need to be hauled from Earth.

The propellant production having been successfully completed, in 2024 two more boosters lift off the Cape and throw their forty-tonne payloads toward Mars. One of the payloads is an unmanned fuel-factory/ERV just like the one launched in 2022; the other is a habitation module containing a crew of four, a mixture of whole food and dehydrated provisions sufficient for three years, and a pressurized methane/oxygen-driven ground rover. On the way out to Mars, artificial gravity can be provided to the crew by extending a tether between the habitat and the burnt-out booster upper stage and spinning the assembly. Upon arrival, the manned craft drops the tether, aero-brakes, and then lands at the 2022 landing site where a fully fueled ERV and fully characterized and beaconed landing site await it. With the help of such navigational aids, the crew should be able to land right on the spot; but if the landing is off course by tens or even hundreds of kilometers, the crew can still achieve the surface rendezvous by driving over in their rover; if they are off by thousands of kilometers, the second ERV provides a backup. However assuming the landing and rendezvous at site number one is achieved as planned, the second ERV will land several hundred kilometers away to start making propellant for the 2026 mission, which in turn will fly out with an additional ERV to open up Mars landing site number three. Thus every other year two heavy lift boosters are launched, one to land a crew, and the other to prepare a site for the next mission, for an average launch rate of just one booster per year to pursue a continuing program of Mars exploration. This is clearly affordable. In effect, this pioneer approach removes the manned Mars mission from the realm of mega-fantasy and reduces it to practice as a task of comparable difficulty to that faced in launching the Apollo missions to the Moon.

The Mars Direct surface base. The tuna can-shaped habitation module is on the left, and the ERV is at right. (Painting by Robert Murray)

The crew will stay on the surface for 1.5 years, taking advantage of the mobility afforded by the high-powered, chemically driven ground vehicles to accomplish a great deal of surface exploration. With a twelve- tonne surface fuel stockpile, they have the capability for over twenty-four thousand kilometers' worth of traverse before they leave, giving them the kind of mobility necessary to conduct a serious search for evidence of past or present life on Mars—an investigation key to revealing whether life is a phenomenon unique to Earth, or something that is happening throughout the universe. Since no one has been left in orbit, the entire crew will have available to them the natural gravity and protection against cosmic rays and solar radiation afforded by the Martian environment, which data from the *Curiosity* rover has shown to be no more hazardous from a radiological point of view than low Earth orbit. Thus there will not be the strong driver for a quick return to Earth that plagues conventional Mars mission plans based upon orbiting mother ships with small landing parties.

At the conclusion of their stay, the crew returns to Earth in a direct flight from the Martian surface in the ERV. As the series of missions progress, a string of small bases are left behind on the Martian surface, opening up broad stretches of territory to human cognizance.

Part 3:

Our Current Predicament

Such are the limitless possibilities potentially within our reach. Yet the reality is that America's human spaceflight program is now adrift. The Space Shuttle has made its final flight, and the Obama administration has no coherent plan as to what to do next. Instead, as matters stand, the United States will waste the next decade spending $100 billion to support a goalless, constituency-driven human spaceflight effort that goes nowhere and accomplishes nothing.

Meanwhile, the robotic exploration program, which has achieved a string of amazing successes over the past sixteen years leading up to the landing of the terrific *Curiosity* rover in August 2012, is currently threatened with budget cuts that could stop it dead in its tracks. This is very ironic, to say the least. The *Curiosity* landing rightly thrilled the world, and the nation's leaders were quick to take bows. "If anyone has been harboring doubts about the status of US leadership in space," the president's science adviser, John P. Holdren, said, "well, there's a one-ton automobile-size piece of American ingenuity. And it's sitting on the surface of Mars right now."

But alas, the *Curiosity* mission is a legacy of the Bush administration, begun by one NASA administrator, Sean O'Keefe, and rammed through to completion over the objections of vocal critics by his gutsy successor, Mike Griffin, who also initiated the *MAVEN* Mars orbiter, scheduled for launching next year. The Obama administration, however, has no plans to continue in like vein. Far from it. It has canceled NASA's plans for joint Mars missions with the Europeans in 2016 and 2018 and is proposing to butcher the program budget.

The figures speak for themselves. The fiscal year 2012 NASA Mars exploration budget was $587 million. The administration cut that to $360.8 million in fiscal year 2013, and is proposing $227.7 million in 2014 and $188.7 million in 2015. These developments pose a grave crisis for any Mars program.

NASA's robotic Mars exploration efforts have been brilliantly successful because, since 1994, they have been approached as a campaign, with probes launched every biennial opportunity, alternating between orbiters and landers. As a result, combined operations have been possible, with orbiters providing communication links and reconnaissance guidance for surface rovers, which in turn can conduct ground-truth investigations of orbital observations. Thus, the great treks of the rovers *Spirit* and *Opportunity*, launched in 2003, were supported from above by *Mars Global Surveyor* (MGS, launched in 1996), *Mars Odyssey* (launched in 2001), and *Mars Reconnaissance Orbiter* (MRO, launched in 2005). But after serving nine years in orbit, MGS is now lost, and if we wait until the 2020s to resume Mars exploration, the rest of the orbiters will be gone as well. Moreover, so will be the experienced teams that created them. Effectively, the whole program will be completely wrecked, and we will have to start again from scratch.

Furthermore, if the administration's cuts are allowed to prevail, we will not only destroy America's Mars exploration program, but derail that of our European allies as well. The 2016 and 2018 missions were planned as a NASA/ESA joint project, with the Europeans contributing over $1 billion to the effort. But if America betrays its commitment, the European supporters of Mars explorations will be left high and dry, and both the missions and the partnership will be lost.

There is no justification for the proposed cuts. The US federal government may be going broke, but it's not because of NASA. Since 2008, federal spending has increased 40 percent, but NASA spending has remained the same. Trillions of dollars of out-of-control entitlement spending cannot be remedied by cuts in NASA, or even in the entire discretionary budget, defense included. Rather, the financial bleeding needs to be staunched where the hole is, and nowhere else.

The proposed cuts caused massive outrage, which was in no way calmed by NASA Administrator Charles Bolden's March 2012 testimony to congress in which he said that the Mars program should be cut because "it had been successful." Some mollification was provided to the Mars scientific community by the August 2012 announcement that NASA's open-for-competition Discovery program had chosen a small Mars geophysical probe called INSIGHT as its 2016 selection. In a further attempt to brighten the picture, NASA Associate Administrator for Science John

Grunsfeld announced in December 2012 that NASA would send another Curiosity-model rover to Mars in 2020. However as no increased funding came along with this message, and as the 2020 target date was clearly chosen to avoid the need for the administration to actually provide any serious cash, the reality of this commitment may be viewed with some skepticism.

The Mars program is not being derailed to make funds available for future missions to other planets. In fact, there is no money in the Obama OMB plan to fund any of them, either.

In any case, cost is not the issue. With the Europeans putting up their share, a matching $1 billion contribution from NASA spread over the next six years would be sufficient to fund both the 2016 and 2018 missions at a level of a billion dollars each. This would require less than 1 percent of NASA's current budget. There is no excuse for not doing this. Indeed, what is truly remarkable about the Obama administration's NASA management is that it has managed to wreck both the human spaceflight program and the robotic planetary exploration effort without saving any money. In 2008, NASA spending was $17.4 billion; the 2013 budget is $17.7 billion. Yet in 2008, NASA was running an active space shuttle program, preparing for the critical mission to save the Hubble Space Telescope, developing systems for returning astronauts to the Moon by 2019, building the *Curiosity* and *MAVEN* Mars probes, and planning an orbiter for Jupiter's moon Europa. Today the shuttles are gone, the Moon program is gone, and this decade's primary planned post-*MAVEN* Mars and Jupiter probes are gone—all without saving a nickel. In terms of damage done per dollar cut, it may be a world record. But it gets worse.

Mars Sample Return the Hard Way

In response to the outrage over its cancelation of the 2016 and 2018 Mars missions, in the summer of 2012, the administration ordered a major "rethink" of NASA's plans for continued robotic exploration of the Red Planet. Reporting back in October 2012, the agency bureaucracy said that the mission of returning a small sample of material (known as the Mars Sample Return, or MSR) from the Red Planet should be the primary goal of the robotic Mars exploration program. That was not particularly remarkable. What was remarkable, however, was the unprecedented and incredibly unnecessary complexity of the plan proposed for achieving such an objective.

It may well be asked whether a sample return would be the best way to pursue the robotic scientific exploration of Mars within the budget of the Mars exploration program run by NASA's planetary exploration directorate. That is an issue over which reasonable people may, and do, differ. It is certainly possible to propose alternative robotic mission sets consisting of assortments of orbiters, rovers, aircraft, or surface networks that might produce a greater science return than the MSR mission much sooner, especially in view of the fact that human explorers could return hundreds of times the amount of samples, selected far more wisely, from thousands of times the candidate rocks, than an MSR mission. However, that said, if the scientific community really believes that a robotic Mars sample is so valuable that it is worth sacrificing all the other kinds of science they could do with their cash, then it is imperative that NASA develop the most efficient MSR plan, to allow the sample to be obtained as quickly as possible and with the least possible expenditure of funds that could be used for other types of Mars exploration missions.

Unfortunately, however, rather than propose the most cost-effective plan for an MSR mission, NASA has set forth the most convoluted, riskiest, costliest approach ever conceived. The *Curiosity* mission demonstrated a system that can soft-land nine hundred kilograms on the Martian surface. With a nine-hundred-kilogram payload, it is possible to land a complete two-stage Mars ascent vehicle capable of flying a capsule with a one-kilogram sample directly back to Earth, as well as a *Spirit*-class rover to gather the samples for it. But instead of proposing such a straightforward plan, NASA baselined a mission conducted in eight parts, including: a) pre-landing a very large rover to collect and cache samples, b) dispatching a Mars ascent vehicle to Mars and performing a surface rendezvous with the rover or its cache, c) flying the Mars ascent vehicle to Mars's orbit to rendezvous with a solar electric propulsion (SEP) spacecraft, d) flying the SEP spacecraft back to near-Earth interplanetary space, e) building a human-tended space station at Lagrange point L2, just above the far side of the Moon, f) flying astronauts to the

Lagrange point space station, g) dispatching astronauts from the Lagrange point space station to take the sample from the SEP spacecraft and return to the Lagrange point space station, and h) conducting extended studies of the sample in the Lagrange point space station.

The kindest thing that can be said about this quintuple-rendezvous plan is that it is probably the unplanned product of the pathology of bureaucracy rather than the willful madness of any one individual.

For a fifth of its cost, NASA could fly five simple direct return MSR missions, *each* of which would have (at least) five times its chance of mission success. So it's hard to imagine any sane person inventing the proposed plan on purpose.

Clearly, though, the group that drifted into it was attempting to force the MSR mission to provide an apparent excuse for the existence of an assortment of other NASA hobbyhorses. For example, we note that it makes use of an L2 space station and electric propulsion, neither of which is necessary or advantageous for use as part of the MSR mission. So how did these two weird ideas get into the plan?

The idea of building a Lagrange point space station was conceived by NASA's human spaceflight directorate during the summer of 2012 as a pseudo-objective to give their program something to do other than endlessly and pointlessly flying astronauts up and down to the low Earth orbiting International Space Station for the next twenty years. The problem, however, is that an L2 space station would serve no useful purpose whatsoever. We don't need an L2 space station to go back to the Moon. We don't need an L2 space station to go to near-Earth asteroids. We don't need an L2 space station to go to Mars. We don't need an L2 space station for anything. So lacking any other purpose, it was given a role in the MSR mission. But this does not help the MSR mission, which could much more simply just return the samples to Earth, where far better lab facilities are available than could ever be installed at L2. Rather, by imposing the L2 station on MSR as a necessary element of the mission plan, the NASA bureaucrats are inserting a tollbooth blocking the way to the accomplishment of the sample return, while radically increasing mission and program cost, schedule, and risk, and decreasing science return.

The same can be said for forcing the use of electric propulsion on the mission, but as this is part of an even bigger boondoggle barring the way to human Mars exploration, it merits a discussion all of its own.

The VASIMR Hoax

In 2010, President Obama canceled the Bush administration's initiative to return astronauts to the Moon by 2019. Instead, said the president, traveling to near-Earth asteroids and then Mars should be our goal. However, the president said in his speech announcing the new vision at Kennedy Space Center on April 15, 2010, that "critical to deep space exploration will be the development of breakthrough propulsion systems," so we can't actually get ready to go anywhere. But don't worry, said the space administration spokesmen, who claimed that NASA is developing a new breakthrough propulsion system, known as VASIMR, which uniquely will make it possible for astronauts to travel safely and quickly to Mars. We can't go to Mars until we have the revolutionary VASIMR, they said, but just wait, it's on the way, and once it arrives, all things will be possible.

Washington is a city known for its smoke and mirrors, but rarely has such total falsehood been touted as a basis for science policy.

VASIMR, or the Variable Specific Impulse Magnetoplasma Rocket, is not new. Rather, it has been researched at considerable government expense by its inventor, Dr. Franklin Chang-Díaz, a close friend and former Shuttle crewmate of NASA administrator Charles Bolden, for three decades. More importantly, it is neither revolutionary nor particularly promising. Rather, it is just another addition to the family of electric thrusters, which convert electric power to jet thrust, but inferior to the ones we already have.

Existing ion thrusters routinely achieve 70 percent efficiency and have operated successfully both on the test stand and in space for thousands of hours. In contrast, after thirty years of research, the VASIMR has only obtained about 50 percent efficiency in test stand burns of a few seconds' duration, and that is only at high exhaust velocity. When the exhaust velocity is reduced, the efficiency drops in direct proportion. This means that the VASIMR's much-chanted (but always doubtful) claim that it could offer significant mission benefit by trading exhaust velocity for thrust is simply false. In contrast, this capability has been demonstrated by the ion-drive propelled *Dawn* spacecraft currently on its way to an asteroid. Finally, if it is to be used in space, VASIMR will require practical high-temperature superconducting magnets, which do not exist.

But wait, there's more. To achieve his much-repeated claim that VASIMR could enable a thirty-nine-day one-way transit to Mars, Dr. Chang-Díaz posits a nuclear reactor system with a power of two hundred thousand kilowatts and a power-to-mass ratio of one thousand watts per kilogram. In fact, the largest space nuclear reactor ever built, the Soviet TOPAZ, had a power of ten kilowatts and a power-to-mass ratio of ten watts per kilogram. There is thus no basis whatsoever for believing in the feasibility of Dr. Chang-Díaz's fantasy power system.

Space nuclear reactors with powers in the range of fifty to one hundred kilowatts and power-to-mass ratios of twenty to thirty watts per kilogram are feasible, and would be of considerable value in enabling ion-propelled high–data rate probes to the outer solar system, as well as serving as a reliable source of surface power for a Mars base. However, rather than spend its research dollars on such an actually useful technology, the administration has chosen to fund VASIMR.

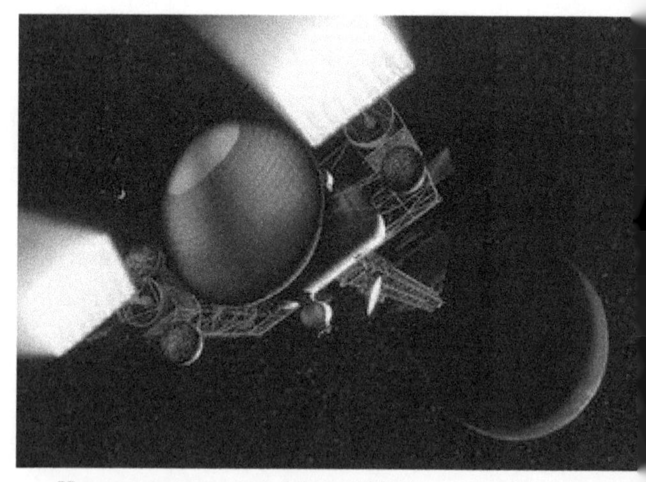

How to never get to Mars: NASA concept for giant nuclear electric spaceship (Art courtesy of NASA)

No electric propulsion system—neither the VASIMR nor its superior ion-drive competitors—can achieve a quick transit to Mars, because the thrust-to-weight ratio of any realistic power system (even without a payload) is much too low. If generous but potentially realistic (someday) numbers are assumed (like, say, fifty watts per kilogram), Dr. Chang-Díaz's hypothetical two hundred thousand–kilowatt nuclear electric spaceship would have a launch mass of 7,700 metric tons, including four thousand tons of very expensive and very radioactive high-technology reactor system hardware requiring maintenance support from a virtual parallel universe of futuristic orbital infrastructure. Yet it would still get to Mars no quicker than the six-month transit executed by the *Mars Odyssey* spacecraft using chemical propulsion in 2001, and which could be readily accomplished by a human crew launched directly to Mars by a heavy-lift booster no more advanced than the (140-ton-to-orbit) Saturn V employed to send astronauts to the

Moon in the 1960s.

That said, the fact that the administration is not making an effort to develop a space nuclear reactor of any kind, let alone the gigantic super-advanced one needed for the VASIMR hyper drive, demonstrates that the program is being conducted on false premises.

So far from enabling a human mission to Mars, VASIMR is primarily useful as a smoke screen for those who wish to avoid embracing such a program. Yet their entire case is disingenuous, because in reality there is no need to develop any faster propulsion system before humans venture to the Red Planet. As noted, the current one-way transit time is six months, exactly the same as a standard crew rotation on the space station. The six-month transit trajectory is actually the best one to use for a human crew, because it provides for a free return orbit, an important safety feature that a faster trajectory would lack. Thus even if we had a truly superior and practical propulsion technology, such as nuclear thermal rockets (which the government is also not developing), we would use its capability to increase the mission payload rather than shorten the transit.

The argument that we must go much faster to avoid cosmic rays is demonstrably false, as proven not only by standard radiation risk analysis (about 1 percent risk of fatal cancer for the fifty-rem, extended-duration, round-trip dose entailed), but by the fact that about a dozen astronauts and cosmonauts have already received such a cumulative cosmic ray dose during repeated flights on the ISS or Mir and, as expected, none of them have evidenced any radiological health effects. (Cosmic ray dose rates on the ISS are fully half of those in interplanetary space, because the Earth blocks out half the sky. The Earth's magnetic field does not shield effectively against cosmic rays. As a result, over the next ten years, ISS crews will receive the same number of person-rems of cosmic radiation as would have been received by five crews of equal size flying to Mars and back over the same period.) As for avoiding zero-gravity deconditioning, the practical answer is to simply prevent it entirely by rotating the spacecraft to provide artificial gravity, rather than waste decades and vast sums in a futile effort to develop warp drive.

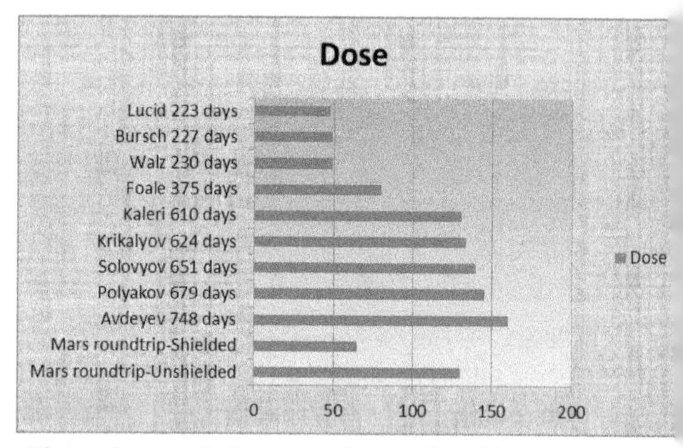

Through extended stays onboard the ISS and Mir, a number of astronauts and cosmonauts have already received cosmic ray doses comparable to a trip to Mars. All survived. (Doses in rems.)

NASA has spent a lot on VASIMR, but its real cost is not the tens of millions spent on the thruster, but the tens of billions that will be wasted as the human spaceflight program is kept mired in Earth orbit for the indefinite future, accomplishing nothing while waiting for the false vision to materialize. And now the constituency created and funded within NASA around the conceit that electric propulsion is on the critical path to human Mars exploration has imposed itself as a tollbooth blocking the path to sample return as well. But far from rejecting this imposition out of hand, those at the top at NASA headquarters have welcomed it, because by including electric propulsion in the sample return they can pretend that they are using MSR to demonstrate a critical technology for human exploration of Mars, and thus taking an important step toward landing humans on the Red Planet, when in fact they are doing nothing of the kind.

NASA is facing an oncoming fiscal tsunami.

There could never be a worse time for the agency to seek to inflate the cost, stretch the schedule, and minimize the return of its missions. The space program is under the gun, and it needs to deliver the goods. Now, more than ever, if we actually want to get a sample from Mars, we need to employ a plan that does so in the simplest, cheapest, fastest, and most direct fashion possible. Now, more than ever, if the human spaceflight program is to survive, it needs to propose a strategy that achieves real and important goals as quickly and cheaply as possible. Under no circumstances should the agency's leadership be turning its missions into Christmas trees on which to hang all the ornaments in the bureaucracy's narcissistic wish box of useless and costly multidecade delays. Yet that is precisely what they are doing. The question is: Why?

Why NASA Is Failing

If we are to understand NASA's current dementia and prescribe a cure, it is necessary to examine the agency's thought processes.

Over the course of its history, NASA has employed two distinct modes of operation. The first prevailed during the period from 1961 to 1973 and may be called the Apollo Mode. The second has prevailed since 1974 and may be called the Shuttle Mode.

In the Apollo Mode, business is (or was) conducted as follows: First, a destination for human spaceflight is chosen. Then a plan is developed to achieve this objective. Following this, technologies and designs are developed to implement that plan. These designs are then built and the missions are flown.

The Shuttle Mode operates entirely differently.

In this mode, technologies and hardware elements are developed in accord with the wishes of various technical communities. These projects are then justified by arguments that they might prove useful at some time in the future when grand flight projects are initiated.

Contrasting these two approaches, we see that the Apollo Mode is destination-driven, while the Shuttle Mode pretends to be technology-driven, but is actually constituency-driven.

In the Apollo Mode, technology development is done for mission-directed reasons. In the Shuttle Mode, projects are undertaken on behalf of various pressure groups pushing their own favorite technologies and then defended using rationales.

In the Apollo Mode, the space agency's efforts are focused and directed. In the Shuttle Mode, NASA's efforts are random and entropic.

To make this distinction completely clear, a

metaphor may be useful. Imagine two couples, each planning to build their own house. The first couple decides what kind of house they want, hires an architect to design it in detail, and then acquires the appropriate materials to build it. That is the Apollo Mode.

The second couple polls their neighbors each month for different spare house-parts they would like to sell, and buys them all, hoping eventually to accumulate enough stuff to build a house. When their relatives inquire as to why they are accumulating so much junk, they hire an architect to compose a house design that employs all the knickknacks they have purchased. The house is never built, but an excuse is generated to justify each purchase, thereby avoiding embarrassment. That is the Shuttle Mode.

In today's dollars, NASA's average budget from 1961 to 1973 was about $20 billion per year—only 10 percent higher than NASA's current budget. To assess the comparative productivity of the Apollo

Mode with the Shuttle Mode, it is therefore useful to compare NASA's accomplishments during the years 1961–1973 and 2000–2012, as the space agency's total expenditures over these two periods are roughly the same.

Between 1961 and 1973, NASA flew the Mercury, Gemini, Apollo, Skylab, Ranger, Surveyor, and Mariner missions, and did all the development for the Pioneer, Viking, and Voyager missions as well. In addition, the space agency developed hydrogen-oxygen rocket engines, multistage heavy-lift launch vehicles, nuclear rocket engines, space nuclear reactors, radioisotope power generators, spacesuits, in-space life-support systems, orbital rendezvous techniques, soft-landing rocket technologies, interplanetary navigation technology, deep space data transmission techniques, reentry technology, and more.

In addition, such valuable institutional infrastructure as the Cape Canaveral launch complex, the Deep Space tracking network, and the Johnson Space Center were all created in more or less their current form. In contrast, during the period from 2000 to 2012, NASA flew forty shuttle missions, allowing it to twice repair the Hubble Space Telescope and repeatedly visit and add several additions to the International Space Station.

About a dozen interplanetary probes were launched (compared to over forty lunar and planetary probes between 1961 and 1973). Despite innumerable "technology development" programs, no new technologies of any significance were actually developed, and no major operational infrastructure was created.

Comparing these two records, it is difficult to avoid the conclusion that NASA's productivity—both in terms of missions accomplished and technology developed—was vastly greater during its Apollo Mode than during its Shuttle Mode. The Shuttle Mode is hopelessly inefficient because it involves the expenditure of large sums of money without a clear strategic purpose. It is remarkable that the leader of any technical organization would tolerate such a senile mode of operation, but NASA administrators have come to accept it. In fact, during his first two years in office, Sean O'Keefe (the NASA administrator from 2001 until early 2005) explicitly endorsed this state of affairs, repeatedly rebutting critics by saying that "NASA should not be destination-driven." The current administrator, Charles Bolden, has reiterated this, claiming that his current objective-free program actually represents the ideal "flexible path."

Yet ultimately, the blame for this multidecade program of waste cannot be placed solely on NASA's leaders, some of whom, such as former administrator Mike Griffin, have attempted to rectify the situation. Rather, the political class must also accept major responsibility for failing to provide any coherent direction for America's space program—and for demanding more than their share of random projects that do not fit together and do not lead anywhere.

It is this pathology that has crippled the human spaceflight program for the past forty years, and which now threatens to paralyze the robotic exploration program as well.

Advocates of the Shuttle Mode claim that by avoiding the selection of a destination they are developing the technologies that will allow us to go anywhere, anytime. That claim has proven to be untrue. The Shuttle Mode has not gotten us anywhere, and can never get us anywhere. The Apollo Mode got us to the Moon, and it can get us back, or take us to Mars. But leadership is required—and for the last four decades, there has been almost none.

The Advent of SpaceX

There is, however, a bright spot on the horizon in the form of a wave of entrepreneurial activity, most particularly that of the SpaceX company. Founded by PayPal semibillionaire Elon Musk, the advent of this company is living proof of the potential for ideas to drive history.

I first met Elon Musk when he came to a fund-raiser for the Mars Society, a Mars exploration advocacy group that I lead, in San Jose in 2001. It was a $500-per-plate event; Musk gave $5,000. He then took the trouble to travel to Denver a few weeks later to visit me at my company, Pioneer Astronautics, after which he gave $100,000, a donation that allowed us to launch the Mars Desert Research Station in southern Utah. But Musk made clear that, while he supported our efforts, he was a person who acted on his own, not through others, and now that he had made his fortune, the question of what he would do next was open. He wanted to do something that would really matter for the future, and as he saw it, the two chief possibilities were opening the way to Mars or commercializing solar energy. I urged him to devote himself to space. Solar energy might be important, but its commercial potential is obvious, and as soon as its technology advances to the point when it becomes competitive, the Invisible Hand of the free market will mobilize the resources to make it happen. Understanding the significance of making humanity into a multiplanet species, on the other hand, requires a person of vision. Solar energy would happen with him or without him; the founding of a new branch of human civilization on Mars might not.

In the end, he decided to do both (and start an electric car company, too) and proceeded to create SpaceX, a truly remarkable firm. Musk was by no means the first zillionaire to launch a hopefully revolutionary space company. There are at least half a dozen others that I could name. But unlike the other would-be space magnates, Musk did not simply throw an expendable chunk of his fortune into the game; he put the full force of his talent and passion into it. When I met Musk in 2001, he had a good grasp of scientific principles, but knew nothing about rocket engines. When I visited him at his first small factory in Los Angeles in 2005, he knew everything about rocket engines. By the time of my next visit a few years later, he had experienced two straight failures of his first launch vehicle, the Falcon 1, but was determined to push on despite the blows to his finances and reputation. It is this level of commitment that has made all the difference. None of the other billionaire-backed space startups have ever cleared the tower. SpaceX has delivered cargo to the space station and will soon be sending people.

Of course, the established aerospace companies, like Lockheed Martin, could develop systems to reach the space station, too, but not on the same schedule or anything remotely like the same price. Lockheed Martin Astronautics in Denver employs roughly nine times the number of people as SpaceX (15,000 verses 1,700), but the factory workforces of both companies are roughly the same (about 1,300 each). Thus SpaceX's overhead is an order of magnitude less. The reason why this is so is because, for the past several decades, the established aerospace companies have worked under a cost-plus contracting arrangement, where their charges to the government have been regulated to consist of their costs plus an additional percentage taken as profit. Under this arrangement, the more the job costs, the more the company makes (which is why, when I was at Lockheed Martin's predecessor, Martin Marietta, back in the early 1990s, a number of us engineers used to joke that "at Martin Marietta, overhead is our most important product"). In contrast, at SpaceX, initially all—and still a significant fraction today—of the funds spent have been Musk's. In short, SpaceX spends money like it is its own—because much of it is.

Operating thus, SpaceX was able to field the medium lift Falcon 9 (ten metric tons to orbit) launch vehicle for a total development cost of $300 million, which is about a tenth the conventional NASA/aerospace industry estimate for such an accomplishment. They were also able to develop the human-capable Dragon capsule in just four years for a further $300 million, while Lockheed Martin and NASA are struggling together to get the development of their crew capsule (known sequentially as the CEV, Orion, and MPCV) done in a decade at an estimated cost on the order of $10 billion.

And now, SpaceX has announced that by 2014 it intends to field the Falcon Heavy, capable of delivering fifty-three metric tons to orbit, doing so entirely on its own money. While not a true heavy-lift booster, this would still be twice the capacity of any rocket now flying, and poses a thrilling question:

Could we use such a system to reach Mars in this decade?

I believe the answer is yes.

Dragon Direct

Here's how it could be done. The SpaceX Falcon Heavy will have a launch capacity of fifty-three metric tons to low Earth orbit. This means that if a conventional hydrogen/oxygen chemical rocket upper stage were added, it would have the capability of sending 17.5 tons on a trajectory to Mars, placing fourteen tons in Mars orbit, or landing eleven tons on the Martian surface. The SpaceX Dragon crew capsule has a mass of about eight tons. While its current intended mission is to ferry up to seven astronauts to the International Space Station, the Dragon's heat shield system is overdesigned and is capable of withstanding reentry not just from Earth orbit, but from interplanetary trajectories. It's rather small for an interplanetary spaceship, but it is designed for multiyear life, and if we cut its crew from seven to two, it should be able to carry supplies enough for a pair of astronauts, with ample additional living space provided by employing a six-meter-diameter by eight-meter-long two-deck inflatable extension.

Using these basic tools, a Mars mission could be done utilizing three Falcon Heavy launches, employing a three-launch variation of Mars Direct known as the Semi-Direct plan. One launch would deliver to Mars orbit an unmanned Dragon capsule with a methane/oxygen chemical rocket stage of sufficient power to drive it back to Earth. This is the Earth Return Vehicle (ERV). A second launch would deliver to the Martian surface an eleven-ton payload consisting of a Mars Ascent Vehicle (MAV) employing a single methane/oxygen rocket propulsion stage, a small automated chemical reactor system, three tons of surface exploration gear, and a ten-kilowatt power supply, which could be either nuclear or solar. The MAV would land with its propellant tanks filled with 2.6 tons of methane, but without the nine tons of liquid oxygen required to burn it. This oxygen could be made over a five-hundred-day period by using the chemical reactor to break down the carbon dioxide that composes 95 percent of the Martian atmosphere. Since the reactor and the power system together only weigh about two tons, using such technology to generate the required oxygen in situ rather than transporting it saves a great deal of mass, and offers the further benefit of providing copious power and unlimited oxygen to the crew once they arrive. Combined, the 11.6 tons of methane/oxygen propellant is sufficient to deliver a two-ton crew cabin (equal in dry mass to the lunar ascent vehicle used during the Apollo missions) from the Martian surface to high Mars orbit, where it can rendezvous with the ERV.

The Dragon and an inflatable extension could create an ample artificial gravity habitation module.
(Painting by Michael Carroll)

Once these elements are in place, the third launch would occur, which would send a Dragon capsule with a crew of two astronauts on a direct trajectory for Mars. The capsule would carry 2,500 kilograms of consumables, sufficient, if water and oxygen recycling systems are employed, to support the two-person crew for up to three years. Given the available payload capacity, a light ground vehicle and several hundred kilograms of science instruments could be taken along as well. Because a six-meter-diameter, eight-meter-long inflatable Kevlar expansion unit need only have a mass of about two hundred kilograms, an extra inflatable could be brought along to deploy on the surface, should it prove impractical to retract and reuse the extension unit employed in space. As in the Mars Direct plan, artificial gravity could be provided by extending a tether between the burned-out trans-Mars injection stage and the Dragon/habitat module and rotating the combination.

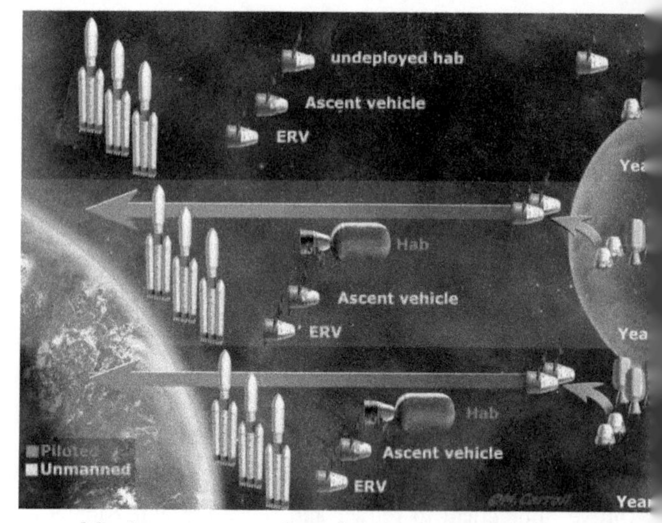

Mission sequence chart for the Dragon Direct plan. Every two years, three Falcon Heavies are launched, sending an Ascent Vehicle, an Earth Return Vehicle, and a piloted habitat. (Painting by Michael Carroll)

The crew would take six months to reach Mars, after which they would land their Dragon capsule near the MAV. They would then spend the next year and a half exploring Mars, using their ground vehicle for mobility and the Dragon/inflatable combination as their home and laboratory (with the Dragon acting as the airlock for the inflatable habitat above it). At the end of their eighteen-month surface stay, the crew would transfer to the MAV, take off, and rendezvous with the ERV. This craft would then take them on a six-month flight back to Earth, whereupon it would enter the atmosphere and splash down to an ocean landing.

There is nothing in this plan that is beyond our current level of technology. Nor would the costs be excessive. Falcon Heavy launches are priced at about $100 million each, and Dragons are even cheaper. Adopting such an approach, we could send human expeditions to Mars at less than the mission cost recently required to launch a Space Shuttle flight.

What is required, however, is a much more rational and courageous approach to decision-making than that followed by the current space policy bureaucracy.

Which brings us to our next topic.

The Question of Risk

> If we could put a man on the Moon, why can't we put a man on the Moon?!
>
> —anonymous frustrated NASA manager, 1990

There is another issue that needs to be dealt with if we are to make it to Mars, or anywhere else in space, and that is the question of risk.

Consider this: Starting with near-zero space capability in 1961, NASA's Apollo program put men on our companion world in eight years. Yet despite vastly superior technology and hundreds of billions of dollars of expenditures, the agency has been unable to send anyone farther than low Earth orbit ever since.

Why? To those of us who have worked in the space program for the past several decades, the answer is clear: because we insist that our astronauts be as safe as possible.

Achieving safety is certainly desirable and merits significant expenditure. But there is a potentially unlimited set of testing procedures, precursor missions, technological improvements, and other protective measures that could, in principle, be implemented before allowing human beings to attempt flights to other worlds. Were we to attempt all of them, we would obtain a human spaceflight program of infinite cost and zero accomplishment. Indeed, in recent years, the trend has gone in precisely such a direction, with NASA's manned spaceflight effort costing more and more to accomplish less and less. If it is to achieve anything, we have to find some way to strike a balance between human life and mission accomplishment.

Put in bureaucratese, what we need is a quantitative criterion to assess what constitutes a rational expenditure to avert astronaut risk. Or, in plain English, we need to answer the following question: *How much is an astronaut's life worth?*

The Worth of an Astronaut

There are three measures that, taken together, can be used to assign a value to the life of an astronaut. First, as a human being, whose life is intrinsically precious. Second, as trained personnel, in whose development the taxpayers have made a significant investment.

And third, as a potentially critical asset to the mission he or she is conducting.

The life of an astronaut is intrinsically precious, but no more than that of anyone else. If we are to assess the rationality of government expenditures to protect the lives of astronauts considered simply as human beings, we need to compare such expenditures to those spent to save the lives of other citizens.

Dr. John D. Graham and his colleagues at the Harvard Center for Risk Analysis have examined precisely this question as it applies to several hundred other government programs. Simplifying their results somewhat for our present purposes, the Harvard researchers found that the median cost for lifesaving expenditures and regulations by the US government in the health care, residential, transportation, and occupational areas ranged from about $0.8 to $2.7 million spent per life saved. The only marked exception to this pattern occurs in the area of environmental health protection (such as the superfund program), which costs about $160 million per life saved, a form of inefficiency that the Harvard researchers term "statistical murder," since if the funds so expended had been used in a more cost-effective manner, many thousands of additional lives per year could have been saved. In fact, in order to avoid such deadly waste, the US Department of Transportation has a policy of rejecting any proposed safety expenditure costing more than $3 million per life saved. This therefore may be taken as a high-end estimate for the value of the life of an American citizen as defined by the US government.

However, astronauts are not just anyone. They are highly trained personnel in whom an investment of several tens of millions of dollars each has been made in order to prepare them to accomplish their missions. The exact figure varies from astronaut to astronaut, as some (such as those who are former fighter pilots) have received much more training than others. Let us therefore err toward the high side and assign a value of $50 million per astronaut, including intrinsic worth and training investment, but setting to one side for the moment their value as critical assets to the mission they are flying.

Looking at the matter this way can provide some useful guidance for weighing risk against expenditure in the human spaceflight program. By way of illustration, let us consider several recent examples.

Hubble Repair: In January 2004, then NASA administrator Sean O'Keefe announced that he was canceling the agency's planned Shuttle mission to save, repair, and upgrade the Hubble Space Telescope. According to Mr. O'Keefe, the January 2003 loss of the Space Shuttle *Columbia* showed how risky such flights actually were, and therefore, as a responsible government official, he could not authorize such a perilous venture. The Hubble Space Telescope is a unique astronomical observatory that, aside from its world-historic value to science (it discovered that the expansion of the universe is accelerating, thereby revealing the existence of a previously unsuspected fundamental physical force), represents a cash investment of about $4 billion on the part of American taxpayers. To be conservative, let us assume that all the safety improvements undertaken after the *Columbia* accident accomplished absolutely nothing, so that the Shuttle reliability was just the 98 percent demonstrated up until that time (123 successful flights out of 125). By the above analysis, the seven-person crew of the Shuttle may be assigned a value of $350 million, to which we must add the replacement cost of the Shuttle orbiter itself, which was about $3 billion. This $3.35 billion is what would be placed at a 2 percent risk in order to accomplish the mission. Two percent of $3.35 billion is $67 million. Comparing this $67 million risk or insurance cost against the $4 billion value represented by Hubble, we can see that Administrator

O'Keefe's argument that Hubble should be abandoned in order to avoid mission risk was completely irrational.

As a sanity check on the above argument, let us compare an analogous situation outside of the space program. Four billion dollars is also the cost of a brand-new, nuclear-powered aircraft carrier. Consider the case of the captain of such a vessel who allowed his ship to sink rather than allow seven volunteers to attempt a repair, on the grounds that the odds in their favor were only fifty to one. Such an officer would be court-martialed and regarded with universal contempt both by his brother officers and society at large. So in this example, the results of our quantitative approach based on risk analysis cohere well with intuition and the generally accepted values of our society.

The Shuttle Stand-down: Following the *Columbia* accident, the Shuttle program was grounded for more than two years, and its flight rate severely limited thereafter. In its last few years of operation, the Shuttle program cost about $4 billion per year and, when conducted in its normal pre-*Columbia* fashion, could be expected to be able to launch about five flights per year. Therefore, while the program was properly underway, American taxpayers were paying for five Shuttle missions annually at a cost of $800 million each. Let us stipulate that these missions were actually worth this much money (if they were not, then the Shuttle program should have been canceled regardless of risk). In that case, each Shuttle mission cut from the yearly manifest of five represented a loss to the taxpayers of $800 million, as they were paying for a mission they were not getting. Once again, comparing this $800 million loss to the $67 million risk associated with the flight, we find no rationale for holding back.

Put simply, if NASA takes $4 billion per year to fly humans in space, it really has to fly them. Four billion per year, if spent on other government lifesaving efforts, would result (at a mean of $2 million per life) in the saving of roughly two thousand lives. This is the sacrifice that the nation is making in order to allow NASA to run a human spaceflight program. In the face of such sacrifice, results are required.

Human Mars Exploration: Mars is key to humanity's future in space. It is the closest planet that has all the resources needed to support life and technological civilization. Its complexity uniquely demands the skills of human explorers, who will pave the way for human settlers. It is, therefore, the proper goal for NASA's human spaceflight program, and, in its public statements, the space agency nominally embraces it as such. However, according to NASA, before it attempts such missions, it must act to minimize risk by conducting a variety of preparatory programs, including the now-ended Shuttle program, the continuing International Space Station program, a variety of robotic probes, the construction of a Lagrange point space station, a set of near-Earth asteroid expeditions, the construction of a lunar base, missions to the Martian moons, and the development of an amusing assortment of other allegedly valuable orbital infrastructure projects and advanced propulsion systems.

Discounting the probes, which don't cost much and actually are quite useful, the rest of this agenda comes with a price tag on the order of half a trillion dollars and a delay in mission accomplishment by half a century. NASA's Apollo-era leadership wanted to send men to Mars by 1981. Their plan was canned in favor of the Shuttle, the ISS, and an extended program of learning how to live and work in low Earth orbit before we venture farther.

It would have been unquestionably risky to attempt a Mars mission in the 1980s, just as it was to reach for the Moon in the 1960s. But even if we ignore the fact that the multidecade preparatory exercise adopted as an alternative to real space exploration has already cost the lives of fourteen astronauts—and will almost certainly cost more as it endlessly continues—it must be asked: How rational is it to spend such huge sums in order to marginally reduce risk to the crew of the perpetually deferred Mars One?

Let's do the math. It's true that nearly anything we do in space will provide experience that will reduce the risk to subsequent missions, but by how much? Suppose that by doing one of the aforementioned intermediate activities, say running the Space Station program for another ten years, we can increase the probability of success of the first expedition to Mars from 90 percent to 95 percent. Assume that the extended Space Station program costs $50 billion, that we disregard its own risk, and that the crew of the first Mars mission consists of five people. Cutting the risk to five people by 5 percent each is equivalent to saving 25 percent of one human life. At a cost of $50 billion, this would work out to $200 billion per life saved, a humanitarian effort a hundred thousand times less efficient than the standard set by the federal highway department, entailing considerable risk of its own, and done at the expense of an additional decade of delay in achievement of the primary mission.

Such an approach makes no sense at all.

The Mission Comes First

The contrast between NASA's current attitude to risk and that of explorers in previous ages of exploration is marked. Neither Columbus nor Lewis and Clark would have imagined demanding 99.999 percent safety assurances as a precondition for launching their expeditions. Indeed, it is quite clear that had such a standard been required, no human exploration voyages could ever have been done. For those courageous souls who sought and found the paths that took our species from its ancestral home in the Kenyan rift valley to every continent and clime of the globe, it was enough that the game was worth the candle and that they had a fighting chance to win it. Perhaps, should a true private entrepreneurial space sector emerge, its captains will take a similar heroic stance. But for now, speaking realistically, it must be said that while SpaceX and the rest may offer the prospect of substantial reductions in cost to the NASA exploration program, they remain vendors to that program, which must supply the funds and therefore call the shots. This makes the question of risk a matter of public policy.

Am I saying that we should just bull ahead, regardless of the risk? No. What I am saying is that *in space exploration, the top priority must not be human safety, but mission success.* These sound like the same thing, but they are not. Let me explain the difference by means of an example.

Suppose you were the manager of a program to send robotic rovers to Mars. You have a fixed budget and two options on how to spend it. The first option is to spend half the money on development and testing, and the rest on manufacturing and flight operations. If you take this choice, you get two rovers, each with a success probability of 90 percent. The other option is to spend three quarters of the budget on development and testing, leaving a quarter for the actual mission. If you do it this way, you only get one rover, but it has a success probability of 95 percent. Which option should you choose?

The right answer is to go for the two-rover plan, because if you do it that way, you will have a 99 percent probability of succeeding with at least one rover, and an 81 percent probability of getting *two* successful rovers—an outcome that is not even possible with the other approach. This being a robotic mission, with no lives at stake, that's all clear enough. But what if we were talking about a human mission? What would be the right choice then? My answer would be the same, because with tens of billions of dollars that could go to meet all kinds of pressing human needs elsewhere being entrusted to the space agency, its first obligation must be to deliver and get the job done. Otherwise, all the sacrifice that treasure represents would be for nothing.

Of course, if the choice were between two missions, one with two flights each with just a 10 percent success probability, and one with a single flight with a 90 percent chance, the correct answer would be different. The point here is not the particular numbers assumed or the specific answers cited. The point is that there is a methodology, well established in other fields of endeavor, that can allow us to assess the rationality of risk reduction expenditures in the human spaceflight program. If NASA disagrees with my suggested assignment of $50 million to the life of an astronaut, it should propose its own figure, substantiate it, and then subject its proposed plan of action to quantitative cost-benefit analysis based upon that assessment. But it needs to be a finite number, for to set an infinite value on the life of an astronaut is to set both the goals of the space exploration effort and the needs of the rest of humanity at naught.

This may seem like a harsh approach. But the many billions being spent on the human spaceflight program are not being spent for the safety of the astronauts. They could be safe enough if they stayed home. It is being spent to open the space frontier. Human spaceflight vehicles are not amusement park rides. They are daring ships of exploration that need to sail in harm's way if they are to accomplish a mission critical to the human future. That mission needs to come first.

Part 4:

The Ultimate Stakes

We believe that free labor, that free thought, have enslaved the forces of nature, and made them work for man. We make the old attraction of gravitation work for us; we make the lightning do our errands; we make steam hammer and fashion what we need. . . . The wand of progress touches the auction-block, the slave-pen, the whipping-post, and we see homes and firesides and schoolhouses and books, and where all was want and crime and cruelty and fear, we see the faces of the free.

—Colonel Robert G. Ingersoll,
Indianapolis speech, 1876

The law of existence prescribes uninterrupted killing, so that the better may live.

—Adolf Hitler, 1941

Will the human future look like *Star Trek* or *Soylent Green*? Is the human prospect unlimited or limited? Is our frontier open or closed? Will we see a future where humanity will choose to do away with want and crime and cruelty and fear, or continue with uninterrupted killing?

These are very important questions, and their answers are a matter of great concern, not only for those who live in distant centuries, but for those alive today. Currently, the limited-resource view is most fashionable among futurists. But if they are right, then human freedoms must be curtailed. Furthermore, world war and genocide would be inevitable, for if the belief persists that there is only so much to go around, then the haves and the want-to-haves are going to have to duke it out, the only question being when.

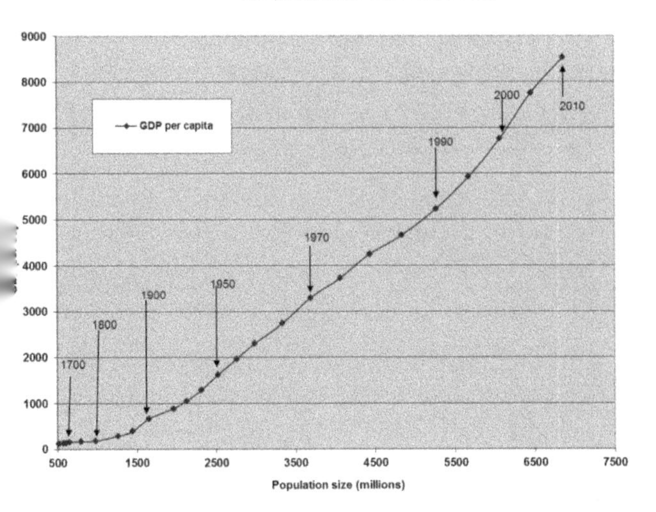

Contrary to Malthus's theory, human global well-being has increased with population size, and at an accelerating rate.

It is notable that the past predictions of resource-depletion theorists have been proven universally wrong. For example, two hundred years ago, the English economist Thomas Malthus set forth the proposition that population growth must always outrun production as a fundamental law of nature. In fact, over the two centuries since, world population has risen sevenfold while inflation-adjusted global gross domestic product per capita has increased by a factor of fifty, and absolute total GDP by a factor of 350.

Indeed, it is clear that the Malthusian argument is fundamentally false, because resources are a function of technology, and the more people there are, and the higher their living standard, the more inventors, and thus inventions, there will be—and the faster the resource base will expand. Nevertheless, so long as humanity is limited to one planet, the arguments of the Malthusians have the appearance of self-evident truth, and their triumph can only have the most catastrophic results.

Once again, to be clear, the issue is not whether space resources will be made available to Earth in the proximate future. Rather it is how we, in the present, conceive the nature of our situation in the future. Nazi Germany had no need for expanded living space. Germany today is a much smaller country than the Third Reich, with a significantly higher population, yet Germans today live much better than they did when Hitler took power. So in fact, the Nazi attempt to depopulate Eastern Europe was completely insane, from not only a moral, but also a practical standpoint. Yet driven on by their false zero-sum beliefs, they tried anyway.

If it is allowed to prevail in the twenty-first century, it will have even more horrific consequences. For example, there are those who point to the fact that Americans are 4 percent of the world's population, yet use 25 percent of the world's oil. If you were a member of the Chinese leadership and you believed in the limited resource view (as many do—witness the brutal one-child policy), what does this imply you should attempt to do to the

United States?

On the other hand, there are those in the United States who cry with alarm at the rising economy and concomitant growing resource consumption of China.

There were no valid reasons for the first two World Wars, and there is no valid reason for a third. But there could well be one if zero-sum ideology prevails.

There is no scientific foundation supporting these motives for conflict. On the contrary, it is precisely because of the freedom and affluence of the United States that American citizens have been able to invent most of the technologies that have allowed China and so many other countries to lift themselves out of poverty. And should China (approaching a population of 1.5 billion) develop to the point where its per-capita rate of invention mirrors that of the United States—with 4 percent of the world's population producing 50 percent of the world's inventions—the entire human race would benefit enormously. Yet that is not how people see it, or are being led to see it by those who should know better.

Rather, people are being bombarded on all sides with propaganda not only by those seeking trade wars or preparations for resource wars, but by those who, portraying humanity as a horde of vermin endangering the natural order, wish to use Malthusian ideology as justification for suppressing freedom. "The Earth has cancer and the cancer is man," proclaims the elite Club of Rome in one of its manifestos. This mode of thinking has clear implications. One does not provide liberty to vermin. One does not seek to advance the cause of a cancer.

If the twenty-first century is to be one of peace, prosperity, hope, and freedom, a definitive and massively convincing refutation of these pernicious ideas is called for—one that will forever tear down the walls of the mental prison these ideas would create for humanity.

Ideas have consequences. Humanity today faces

a choice between two very different sets of ideas, based on two very different visions of the future. On the one side stands the antihumanist view, which, with complete disregard for its repeated prior refutations, continues to postulate a world of limited supplies, whose fixed constraints demand ever-tighter controls upon human aspirations. On the other side stand those who believe in the power of unfettered creativity to invent unbounded resources and so, rather than regret human freedom, demand it as our birthright. The contest between these two outlooks will determine our fate.

If the idea is accepted that the world's resources are fixed with only so much to go around, then each new life is unwelcome, each unregulated act or thought is a menace, every person is fundamentally the enemy of every other person, and each race or nation is the enemy of every other race or nation. The ultimate outcome of such a worldview can only be enforced stagnation, tyranny, war, and genocide.

Only in a world of unlimited resources can all men be brothers.

On the other hand, if it is understood that unfettered creativity can open unbounded resources, then each new life is a gift, every race or nation is fundamentally the friend of every other race or nation, and the central purpose of government must not be to restrict human freedom, but to defend and enhance it at all costs.

That is why we need to open the space frontier. That is why we must take on the challenge of launching a new, dynamic, pioneering branch of human civilization on Mars—whose optimistic, impossibility-defying spirit will continue to break barriers and point the way to the incredible plentitude of possibilities that urge us to write our daring, brilliant future among the vast reaches of the stars. We need to show for all to see what the great Italian Renaissance humanist Giordano Bruno boldly proclaimed, that "there are no ends, limits, or walls that can bar us or ban us from the infinite multitude of things."

Bruno was burned at the stake by the Inquisition for his daring, but fortunately others stepped up to carry the banner of reason, freedom, and dignity forward to victory in his day. So we must do in ours.

And that is why we must begin by taking on the challenge of Mars. For in doing so, we make the most forceful statement possible that we are living not at the end of history, but at the beginning of history; that we believe in freedom and not regimentation, in progress and not stasis, in love rather than hate, in life rather than death, and in hope rather than despair.

Appendix

Founding Declaration of the Mars Society

The time has come for humanity to journey to Mars.

We're ready. Though Mars is distant, we are far better prepared today to send humans to Mars than we were to travel to the Moon at the commencement of the space age. Given the will, we could have our first teams on Mars within a decade.

The reasons for going to Mars are powerful.

We must go for the knowledge of Mars. Our robotic probes have revealed that Mars was once a warm and wet planet, suitable for hosting life's origin. But did it? A search for fossils on the Martian surface or microbes in groundwater below could provide the answer. If found, they would show that the origin of life is not unique to the Earth and, by implication, reveal a universe that is filled with life and probably intelligence as well. From the point of view of learning our true place in the universe, this would be the most important scientific enlightenment since Copernicus.

We must go for the knowledge of Earth. As we begin the twenty-first century, we have evidence that we are changing the Earth's atmosphere and environment in significant ways. It has become a critical matter for us to better understand all aspects of our environment. In this project, comparative planetology is a very powerful tool, a fact already shown by the role Venusian atmospheric studies played in our discovery of the potential threat of global warming by greenhouse gases. Mars, the planet most like Earth, will have even more to teach us about our home world. The knowledge we gain could be key to our survival.

We must go for the challenge. Civilizations, like people, thrive on challenge—and decay without it. The time is past for human societies to use war as a driving stress for technological progress. As the world moves toward unity, we must join together, not in mutual passivity, but in common enterprise, facing outward to embrace a greater and nobler challenge than that which we previously posed to each other. Pioneering Mars will provide such a challenge. Furthermore, a cooperative international exploration of Mars would serve as an example of how the same joint action could work on Earth in other ventures.

We must go for the youth. The spirit of youth demands adventure. A humans-to-Mars program would challenge young people everywhere to develop their minds to participate in the pioneering of a new world. If a Mars program were to inspire just a single extra percent of today's youth to scientific educations, the net result would be tens of millions more scientists, engineers, inventors, medical researchers, and doctors. These people will make innovations that create new industries, find new medical cures, increase income, and benefit the world in innumerable ways to provide a return that will utterly dwarf the expenditures of the Mars program.

We must go for the opportunity. The settling of the Martian New World is an opportunity for a noble experiment in which humanity has another chance to shed old baggage and begin the world anew, carrying forward as much of the best of our heritage as possible and leaving the worst behind. Such chances do not come often and are not to be disdained lightly.

We must go for our humanity. Human beings are more than merely another kind of animal—we are life's messenger. Alone of the creatures of the Earth, we have the ability to continue the work of creation by bringing life to Mars, and Mars to life. In doing so, we shall make a profound statement as to the precious worth of the human race and every member of it.

We must go for the future. Mars is not just a scientific curiosity; it is a world with a surface area equal to all the continents of Earth combined, possessing all the elements that are needed to support not only life, but technological society. It is a New World, filled with history waiting to be made by a new and youthful branch of human civilization that is waiting to be born. We must go to Mars to make that potential a reality. We must go, not for us, but for a people who are yet to be. We must do it for the

Martians.

Believing therefore that the exploration and settlement of Mars is one of the greatest human endeavors possible in our time, we have gathered to found this Mars Society, understanding that even the best ideas for human action are never inevitable, but must be planned, advocated, and achieved by hard work. We call upon all other individuals and organizations of like-minded people to join with us in furthering this great enterprise. No nobler cause has ever been. We shall not rest until it succeeds.

The above declaration was signed and ratified by the seven hundred attendees at the Founding Convention of the Mars Society, held August 13–16, 1998, at the University of Colorado at Boulder, Colorado. If you agree, I invite you to join. Further information is available at www.marssociety.org or by writing the Mars Society, 11111 W. 8th Ave, Unit A, Lakewood, CO, 80215.

About the Author

Dr. Robert Zubrin is president of Pioneer Astronautics, president of the Mars Society, a contributing editor of *The New Atlantis*, and the author of several books, including *The Case for Mars: The Plan to Settle the Red Planet and Why We Must* (Free Press, 1996, 2011); *Entering Space: Creating a Spacefaring Civilization* (Tarcher, 1999); *First Landing* (Ace, 2001); *Mars on Earth: Adventures of Space Pioneers in the High Arctic* (Tarcher, 2003); *Energy Victory: Winning the War on Terror by Breaking Free of Oil* (Prometheus Books, 2007); *How to Live on Mars: A Trusty Guidebook to Surviving and Thriving on the Red Planet* (Three Rivers Press, 2008); and *Merchants of Despair: Radical Environmentalists, Criminal Pseudo-Scientists, and the Fatal Cult of Antihumanism* (Encounter Books, 2012). He lives in Colorado.

CPSIA information can be obtained
at www.ICGtesting.com
Printed in the USA
LVHW052229060619
620461LV00001B/9/P 9 780974 144351